Imagine this...

Stories Inspired by Agriculture
2010

Guadalupe Landeros
Morgan Overholtzer
Zeb Soyffer
Nick O'Brien
Delaney Black
Matt Wright

TWENTY FIVE YEARS

CALIFORNIA FOUNDATION
for **AGRICULTURE** in the **CLASSROOM**

In partnership with

Walmart ⁎

Publisher's Cataloging-in-Publication Data

Landeros, Guadalupe
  Imagine this : stories inspired by agriculture 2010 / Guadalupe Landeros, Morgan Overholtzer, Zeb Soyffer, Nick O'Brien, Delaney Black, Matt Wright.
  p. cm.
  ISBN 978-0-615-44052-1
  Summary: 2010 winning stories from the California Foundation for Agriculture in the Classroom, Imagine this... Story Writing Contest written by third through eighth grade students.

[1. California—Social life and customs—Fiction. 2. Farm life—California—Fiction. 3. Food—Fiction. 4. Short stories. American.] I. Landeros, Guadalupe. II. Overholtzer, Morgan. III. Soyffer, Zeb. IV. O'Brien, Nick. V. Black, Delaney. VI. Wright, Matt. VII. Title.

PZ5 .I326 2011
813.54—dd22   2011922029

California Foundation for
Agriculture in the Classroom
2300 River Plaza Drive
Sacramento, CA 95833
(916) 561-5625 • Fax (916) 561-5697
(800) 700-AITC (2482)
www.LearnAboutAg.org

For Ciara Chiesa
2009 seventh grade state-winning author

Ciara inspired us all with her courage, strength
and love of life, and whose words allowed us a
glimpse in to her beautiful world of gardens,
smiles and rainbows.

Table of Contents

Introduction

For seventeen years the California Foundation for Agriculture in the Classroom has coordinated the *Imagine this...* Story Writing Contest to create agricultural awareness through reading, writing and the arts while meeting Content Standards for California Public Schools. By understanding how and where their food is grown, children and their parents will be inclined to make healthy and nutritious choices at the supermarket, contributing to stronger bodies and sharper minds.

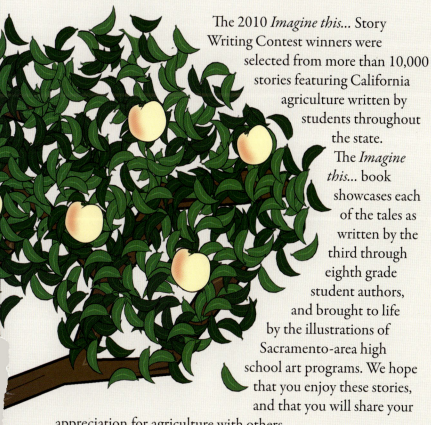

The 2010 *Imagine this...* Story Writing Contest winners were selected from more than 10,000 stories featuring California agriculture written by students throughout the state. The *Imagine this...* book showcases each of the tales as written by the third through eighth grade student authors, and brought to life by the illustrations of Sacramento-area high school art programs. We hope that you enjoy these stories, and that you will share your appreciation for agriculture with others.

Once Upon a Delicious Dream

By Guadalupe Landeros
3rd Grade, Williams Elementary School
Sherrie Taylor Vann, Teacher
Colusa County
Illustrated by Delta High School

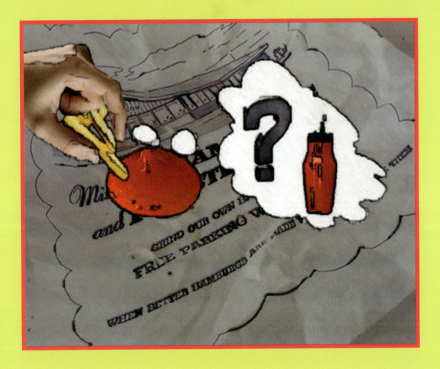

One day at the hamburger stand Lupita was eating some French fries. As she dipped her fry into the pile of bright red ketchup she started to daydream about how it was made.

She had seen lots of trucks piled high with tomatoes **harvested** from the grower's fields driving down the highway headed to a processing plant. Most of the tomatoes arrive at the processing plant and are weighed on a large scale while they are on the trucks. Some of the tomatoes are scooped up by a machine to be **graded**. Lupita thought, "I hope they earn an A!"

Lupita watched to see how they were graded. First, the **inspectors** looked for worms, mold, good color, and things that don't belong with the tomatoes, like leaves, rocks, and sticks. Sometimes things like watermelons sneak in with the tomatoes. The tomatoes are dropped off so the trucks can go back to the fields to pick up more tomatoes. The trucks whizzed by so fast that they blew her hair across her face. "Yikes!" thought Lupita.

When the tomatoes are unloaded they go down a water slide. "Whee! This is fun," said the little tomatoes. Lupita wanted to splash down the waterslide too! They whirled around the waterslide while a camera took pictures of them and looked for bad tomatoes. As they slid in the water, combs swept through the tomatoes and took out any vines or leaves they could reach.

By the time the tomatoes reached the bottom they were clean and shiny from their waterslide. The water goes into a big pond to be reused. "That's good for the planet because it does not waste a lot of water," thought Lupita.

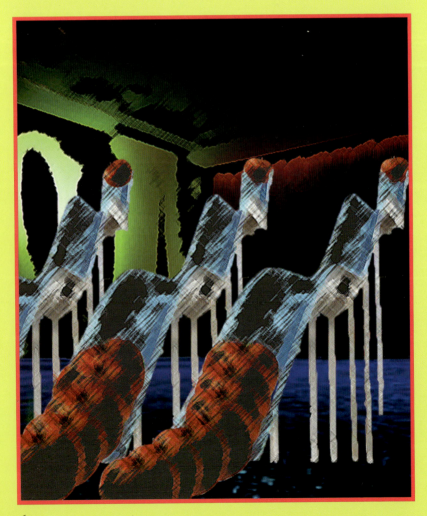

Now that the tomatoes are clean they are cooked at a really hot temperature. They are pushed through screens to take off the skin and take out the seeds. The skins are taken to a farm and fed to pigs. The machine that fills the truck with the skins is like a robot and a person moves it around like a video game. "I'd like to spill some on my teacher's head because she gave me too much homework," thought Lupita, "but she would get really mad!"

After the tomato skins and seeds are removed, the tomato paste has to be cooled down so it can be packed for shipping. A machine cools down the tomato paste instantly by using cool water and a vacuum to suck out the heat. The water is really hot after that so the water goes around a pond to cool off again so it can be reused. Lupita thought, "it would be funny to push my brother, Tony, into the pond because he took some of my French fries!"

Once the paste is cool it gushes through big metal pipes to the packing warehouse. It goes through the pipes into plastic lined bins. The huge, hard plastic bins are filled with 300 gallons of fresh tomato paste. Each bin is sealed so air cannot get in and spoil the paste. "Wow! That is going to make a lot of ketchup for my fries," thought Lupita.

The filled bins are stacked by people driving forklifts while they wait for the delivery trucks. When the trucks are loaded they take the tomato paste to other companies that make things like ketchup or pizza sauce. "Yum, tomatoes make delicious foods!" thought Lupita.

"Finish eating your French fries Lupita because we are all done and ready to go," said Lupita's mom. Lupita's mind came back to her family and her dinner. "OK, OK, I am finished," answered Lupita. She devoured one last bite dripping in red delicious ketchup and went home with her family. 🌿

About the Author

Guadalupe Landeros, age 9

During a school lesson on the top commodities grown in Colusa County, Guadalupe (Lupita) Landeros, a third grader at Williams Elementary School, was particularly interested in processing tomatoes. Although she lives in an agriculturally rich community, Lupita didn't know very much about this popular commodity. While researching and writing her award-winning story, *Once Upon a Delicious Dream*, Lupita was surprised to learn that so many commonly used products are made from tomatoes. Lupita thinks is it important to learn about the things we eat and was excited that part of her research included going on a field trip. Lupita is proud that her hard work paid off and she looks forward to learning and writing more stories about agriculture!

About the Illustrators

Delta High School - River Delta Unified School District
Reed Barnes, A.J. Yarwood, Jeremy Figueira, Octo Ventura,
Mikey Terry, Dillon Stall, Michelle Arceo, Rachale Harmon,
Jonathan Williams, and Eric Davis (not pictured)
Art Instructor: Ian Fullmer

Delta High School's digital media class worked as a team to illustrate the award-winning story, *Once Upon a Delicious Dream.* The students were very familiar with the product they were illustrating but knew very little about the process that tomatoes go through to become ketchup.

After researching fields, trucks, and processing plants, the group broke the story up by page number and each student was assigned a page to design. The students were constantly communicating with each other to make sure that all the pages flowed. They used computers to enhance details and add colors to the finished illustrations.

This experience was new for most of the students and it provided them with a sense of accomplishment to be able to work together to complete the project. During this project, the group also had the opportunity to fine tune their digital photography skills.

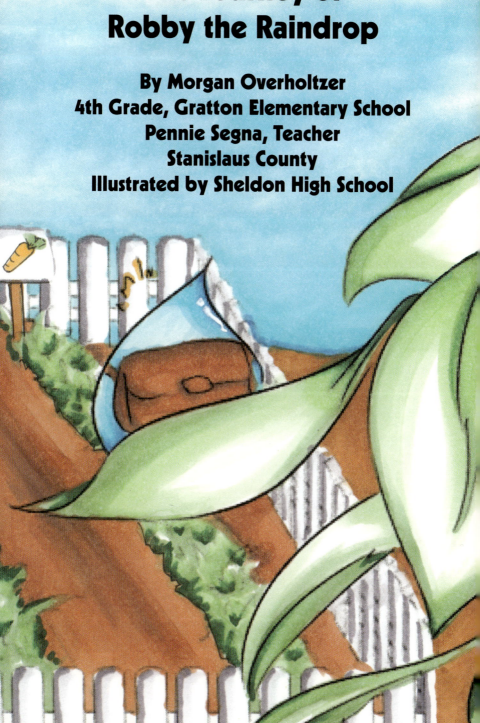

The Journey of
Robby the Raindrop

By Morgan Overholtzer
4th Grade, Gratton Elementary School
Pennie Segna, Teacher
Stanislaus County
Illustrated by Sheldon High School

One sunny day in August, Robby the raindrop was floating along on a cloud with his friends. They were on their way to take water to the dry Central Valley in California. Suddenly the wind started to blow and the cloud got heavier and darker. Robby looked down and said, "I see a lot of peach and almond **orchards**. There is a lot of hay and gardens growing. It looks like they need a lot of water to grow their food. I think I'll jump off here!" So down, down, down Robby fell, passing other clouds and raindrops.

He finally landed on the roof of a house, "Splat," and slid off the edge of a roof top. Robby was dumped right into a garden where he landed on a leaf. Robby the raindrop blinked his eyes and looked around and saw squash, carrots, and cucumbers growing in the garden. When he jumped off of the leaf to look around, he realized that the soil around the plants was very dry. Robby knew the soil needed him because he could put **moisture** into the ground to help plants grow. But, he really didn't realize just how much the soil needed him.

"Help, help," said Robby. "This ground is soaking me up."

He watched as the plant's roots took him in and he started on a journey through the plant's stems until he popped right out of the leaves again. As he was glistening on top of the plant's leaf, he watched sprinklers running. Sprinklers are used to bring up water from the wells deep in the ground to keep the grass healthy and green.

Pretty soon he heard a bubbling noise and looked under the fence to see what it was. The neighbor was irrigating his field, and water was gurgling out of an **irrigation** ditch to keep the almond trees alive. Water was spreading out over the field for as far as he could see! He knew the farmers of California depend on the snow in the winter. When it melts it fills the **dams** and comes rushing down the **canals** during the spring and summer to all the farms.

As he was wondering where to go next, Robby heard a loud "moooo" noise that made him jump! He carefully looked around until he saw a big black and white animal. It was chewing grass and slurping up water. That thing must be a cow, Robby thought. Of course they need a lot of water to make milk! "Wow, they sure are big, beautiful animals," said Robby.

As he turned to go, he slipped and slid down into the gutter by the road. There was so much water gushing down, that he couldn't get his balance. As he zoomed along, he passed the orchard full of trees and the field full of cows. He slid on down the road and fell into the big gutter grate. It was dark and slimy in the pipe that Robby fell into. Soon he was dumped out again, and this time into a river!

As Robby floated along he saw so many interesting things. He passed a family having a picnic by the river and saw the children playing in the water. As he floated on, he watched cows and horses drink water from the river's edge. He also saw bikers and joggers with water bottles, drinking water as they stopped for a rest. People are made up of about 60 percent water, so they need to drink a lot of water every day to keep their bodies hydrated. As Robby the raindrop floated by the big beautiful farms, he realized that everyone was using water!

Wow, I am really important, he thought. Without water like me, plants, animals, and people would not be able to live! So Robby laid his head back and closed his eyes with a happy smile on his face and floated on his way for many miles. It was late in the day when he noticed that the air was changing and he could hear seagulls calling. "I've arrived at the ocean!" Robby exclaimed. He was so excited to be able to float on the ocean waves, but he also knew that he would soon start a new journey. A new water cycle journey, where Robby the raindrop would swiftly **evaporate** and deliver water again to all living things! 🌱

About the Author

Morgan Overholtzer, age 9

Fourth grader Morgan Overholtzer from Gratton Elementary School knew that writing a story for the *Imagine this...* Story Writing Contest was going to be a big job. Morgan is surrounded by agriculture living in Stanislaus County, so she used this opportunity to learn more about the importance of water and to write a story that would help teach others.

While researching, Morgan learned that water is important to the survival of all living things; she learned that our bodies are 60 percent water! Morgan admits the assignment turned out to be more fun than she thought, and she was excited to learn that her story, *The Journey of Robby the Raindrop*, was selected as a state winner!

About the Illustrators

Sheldon High School - Elk Grove Unified School District
Karolyn Chao and Stefanie Serino
Art Instructor: Debbie George

Karolyn Chao and Stefanie Serino were honored to be asked to illustrate the story, *The Journey of Robby the Raindrop*. The two juniors from Sheldon High School were excited to work together on this project since they are good friends and can rely on each other's strengths. The two were able to divide the project up according to their skills in order to ensure the project would get done on time.

Both students had a basic understanding of agriculture before they started the project, they both expanded their knowledge while researching irrigation systems and plant varieties to complete the illustrations. The students are very proud of the work they have done and are excited to see how the students will react to the illustrated story. This project taught Karolyn and Stefanie a lot about time management and meeting deadlines.

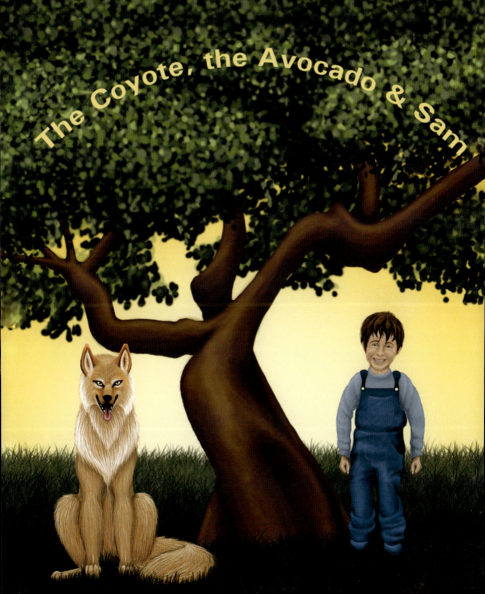

The Coyote, the Avocado & Sam

By Zeb Soyffer
5th Grade, Ekstrand Elementary School
Catherine Rojas, Teacher
Los Angeles County
Illustrated by Woodland High School

One day a little boy named Sam was out by some avocado trees
thinking of how yummy they always taste and decided to climb up
and pick some for himself. Avocados were Sam's absolute favorite
fruit. While he was up in the tree a very large and hungry looking
coyote appeared by the trunk of the tree. Sam grew frightened as he
knew coyotes often ate animals, and sometimes even people! The
coyote asked the boy to come down and play, but Sam knew what
he was up to and said, "No way, I know what you want."

Just then Sam had an idea. He told the coyote of the many health and beauty **benefits** the avocado has including providing 4 percent of the recommended daily allowance of vitamin C in just one avocado. He also explained how they can improve your skin and give your hair- or in the coyote's case, fur- a beautiful silky shine.

While up in the tree, Sam dropped down a tasty snack he had prepared of avocado with a squirt of lemon, topped with a dash of salt and pepper and said to the coyote, "Just try it and see what you think." To the coyote's surprise, it was delicious. He quickly finished it and asked for more.

Sam ended up giving him a dozen or so more of the delicious fruit along with some recipes he said he should try. The coyote's stomach was so full that Sam finally felt safe enough to come down. He told the coyote he could grow them himself and gave him the seeds from the avocados he had eaten and told him how to do it.

One way is to put the pit in a glass of water with a toothpick sticking out of each side so that it doesn't go into the water all the way. Once the avocado pit begins to sprout it should be put into the ground to grow.

Or, if he wanted he could start it in a pot like you would any other type of plant.

He explained that the avocado tree loves plenty of sunshine and a good soak of water, but it's a good idea to let it dry out in between waterings. Sam then told the coyote all about the benefits of being a vegetarian and that the avocado was a perfect food in every way.

The large coyote headed home feeling great, without the usual heaviness he always had after a meal of meat. He felt light and full of energy as he carried a bag full of avocados on his back.

When he entered his cave, his fellow pack of coyotes were mad because he came in without the usual meal of meat. They all stood up and started barking and howling at him in anger.

They were all used to eating small animals, roadkill or anything heavy and meaty, but before they could howl any longer, the large coyote whipped up some guacamole he learned how to make from Sam.

The coyote then watched with delight as his pack of coyotes danced around with joy at the yummy, filling, and tasty treat. They asked where in the world did he find this incredible food and the coyote said from a boy he almost ate!

He then went on to share with them how he intended to grow them with the seeds he had. A few years went by and this former hungry pack of coyotes sent word to the boy Sam to come to their cave for a feast of fresh avocados grown in their very own garden.

When Sam arrived he was in shock at the rows and rows of avocados he saw and immediately thought of how much money they could make if they sold them to the local markets. It turns out that they had so many avocados from their harvest that not only did they make a lot of money, but they wound up buying a bunch of land and growing more than 100 avocado trees.

Before too long they were the most famous avocado growers in the whole state of California and people came from all over the country to go to this special orchard, not only because the avocados were so delicious and full of vitamins but also because they were grown by vegetarian coyotes! Sam was happy because he knew animals and people would no longer be hunted by coyotes and because he could climb trees all day long without worry! 🌿

About the Author

Zeb Soyffer, age 10

Fifth grader Zeb Soyffer from Ekstrand Elementary School has always been encouraged by his parents to use his imagination. He remembers his parents telling him make-believe stories when he was younger. This led to his success as an award-winning author for his story, *The Coyote, the Avocado & Sam.* Zeb and his family have sprouted avocado seeds before, so he decided to use that experience for his story. While doing research for the story, Zeb was surprised to learn that there are eight varieties of avocados grown in California. When he was first given the writing assignment, Zeb was a little nervous because he had never entered a contest like this before. However, he knew that interesting characters and a good conclusion were key when writing a creative story and he is now excited to be a published author.

About the Illustrator

Woodland High School - Woodland Joint Unified School District
Nicole Jackson
Art Instructor: Dawn Abbott

As a senior at Woodland High School, this is Nicole's third year illustrating stories for the *Imagine this...* book. She believes this project has been very beneficial to her and her fellow students, particularly having worked on three different student stories. Through the years Nicole has learned something from each story she has illustrated. This year while illustrating *The Coyote, the Avocado & Sam*, she learned about the benefits of avocados. She has gained an appreciation for agriculture and everything it provides for us on a daily basis. This is Nicole's last year to illustrate a story as she will be graduating later in the year, but she is excited to have all the illustrations to put in her portfolio as she continues her education in art and graphic design.

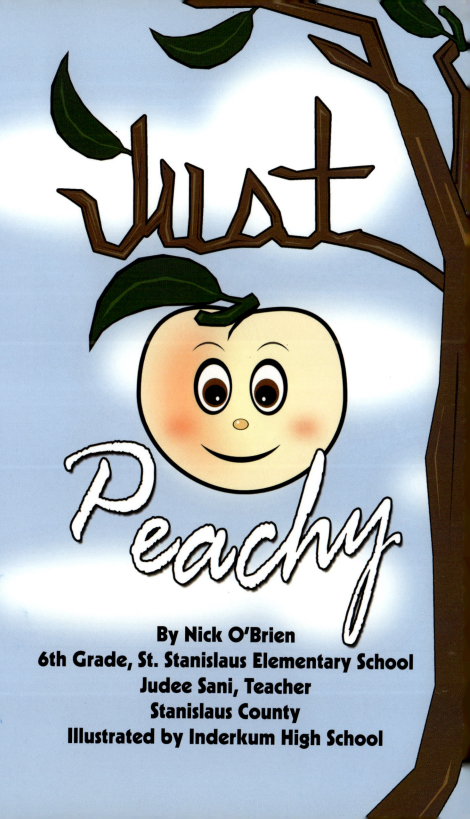

By Nick O'Brien
6th Grade, St. Stanislaus Elementary School
Judee Sani, Teacher
Stanislaus County
Illustrated by Inderkum High School

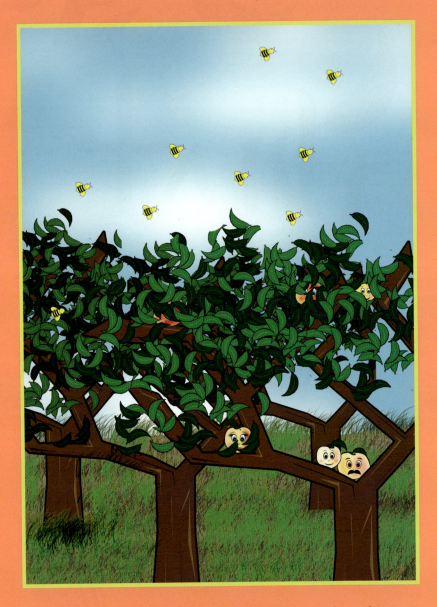

When I was a young peach, my dad was the president of our orchard in Modesto, California. Our orchard was the best in the fruit nation. With the bees flying all over the place shouting and bossing all the bugs around, the orchard was busy and beautiful.

My best friend was another peach five branches up. His name was Johnson and our favorite game was Jump-Branch. Jump-Branch is a game when one peach swings the branch back and forth while the other peach leaps over the branch. It is kind of like jump rope with a branch. My name is Sims and my friend Johnson and I were both **Clingstones**, which makes sense. The two of us were inseparable, very clingy, and did everything together the whole summer.

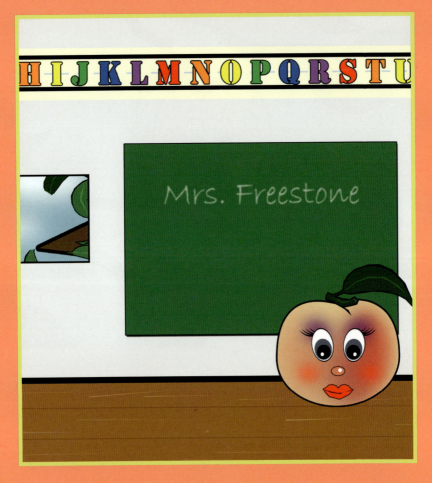

Our school, California Gold Elementary, was located on the second branch. Mrs. Freestone was our teacher and she had bright red fuzz and a bruise on her rear. I always liked Mrs. Freestone; she was sweet. Even though some of us were not **Freestones** like her, she still treated all of the peaches equally. Clingstones were not as popular as the Freestones, but she did not care and taught all of us not to judge a peach by its skin. She never bruised our egos.

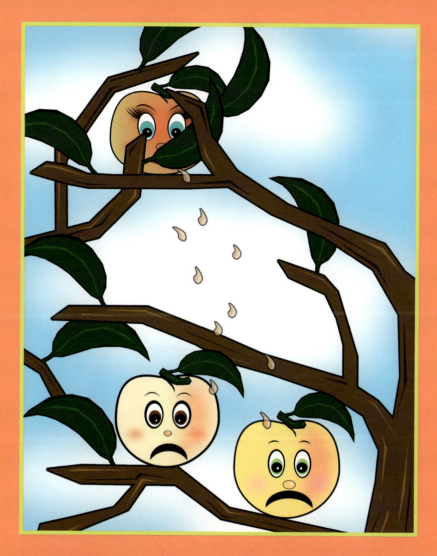

Our next branch neighbor was this old peach named Ms. Faye Elberta and was she ever mean! She thought she was better than everyone because she was a part of the Freestone family and her brand of peach was popular with the eaters. Ms. Elberta was always peeking through the leaves on the tree at my friends and me while we played. Every time we really started having fun, she would make rustling noises in the leaves and then spray us with peach juice. She had a fit when she saw us swinging on the branches kicking up peach fuzz.

Two trees away there was a very special peach named Redhaven. She had the best sunlight in the whole field perched high on the tree and all of the peaches talked about how sweet and perfect she must be soaking all of that sun up every day. My friends and I all dreamed about Redhaven.

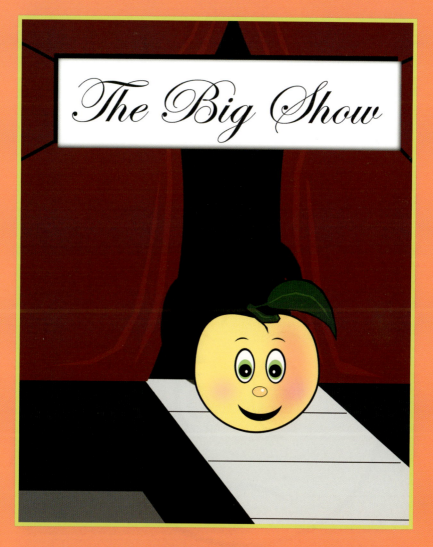

The Big Show

At the end of the California summer, we saw men picking peaches off the branches. After four days of watching the workers picking peaches, they soon were one tree away from our tree. When the men came to our tree, I was hoping that they wouldn't pick anybody I knew. They picked me and everybody else. No one knew where we were going. The workers placed us all into a box, then into a truck, and finally into a factory. We were instructed to all participate in a runway fashion show where we would all parade down the runway belt and hopefully get chosen for the "big show."

Worried that no one would want me, I was very happy when a lady came over and picked me. She placed me in a special pile going to the farmers market. Not all of us got to go to the farmers market, some of my family headed to grocery stores and others to a can of cling peaches. I saw mean old Ms. Elberta going down the fruit cocktail **chutes**. It was sad saying goodbye to everyone and I shed some tears of juice.

My dreams came true the next day. My mom and dad ended up at the farmers market with me. Rolling out of the giant plastic bin with loads of other peaches, I landed right next to Redhaven! Not only was she sweet on the inside, but also on the outside. We chatted for days until a happy little girl picked me and bought me. After shining me up, she took a big bite out of me and was delighted. Enjoying every bite, the girl talked her mom into letting her plant the peach seed in her backyard. Now I am in the ground waiting to grow into a peach tree. See you this spring! 🌱

About the Author

Nick O'Brien, age 12

Nick O'Brien is no stranger to agriculture. Living outside Modesto, surrounded by orchards, made it easy for Nick to select peaches, his favorite fruit, as the topic for his story. A sixth grader at St. Stanislaus Elementary School in Stanislaus County, Nick used his personal experiences watching workers in the fields and observing different varieties of peaches at the farmers market to bring his story, *Just Peachy*, to life. Nick is very happy with the way his story turned out and hopes he was able to make a connection with the readers through his personification of the peaches!

About the Illustrator

Inderkum High School - Natomas Unified School District
Haylee Saelor
Art Instructor: Michelle Townsend

Haylee Saelor felt honored to illustrate *Just Peachy* because she loved the idea of bringing a student's writing to life through art. Haylee is a senior at Inderkum High School and a member of the digital photography class. She used the skills she learned in class to design and complete the illustrations entirely on the computer using Photoshop. Living in the Sacramento Valley, Haylee had seen orchards and had some basic understanding of agriculture, but this project helped her learn much more about the diversity of this important industry. It was a great experience for Haylee to be a part of another student's creative efforts while also improving her own skills.

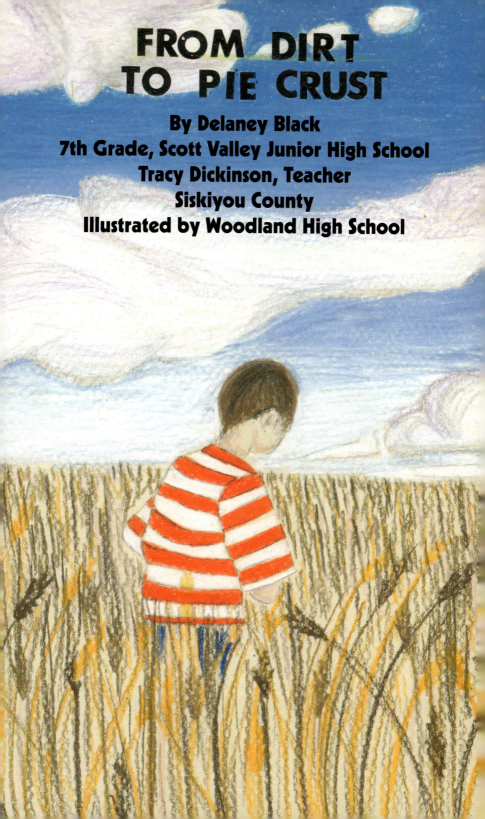

FROM DIRT TO PIE CRUST

By Delaney Black
7th Grade, Scott Valley Junior High School
Tracy Dickinson, Teacher
Siskiyou County
Illustrated by Woodland High School

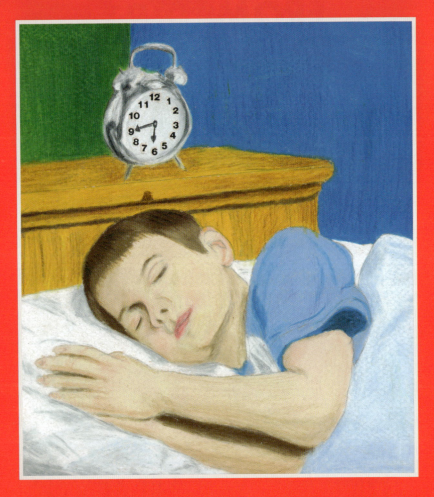

BEEP!! BEEP!! BEEP!! The alarm goes off and two minutes later Dad comes in.

"Get up and get dressed," Dad says as he walks out. 6:45 a.m., I rub my eyes groaning, I want to stay in bed. I get up and get ready, trying to wake myself up while getting dressed.

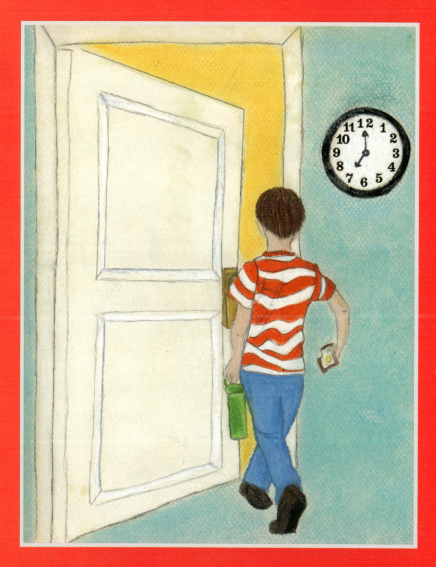

"Hurry up and eat," Dad says. I finish eating while walking out the door promptly at 7:00 a.m. My job is to irrigate the wheat and alfalfa by moving **wheelines** on our family farm. I live on a farm that grows **organic** alfalfa and wheat. We irrigate the wheat until the middle of July. I hurry out to the 4-wheeler and head to the fields with the cool morning air and mosquitoes hitting my face. Every morning, sometimes mid-day, and night, I change water on our family farm.

I turn off the first pipe; wait for it to drain while turning off the others. I move the first, and then walk to the second, and so on. Lastly, I hook them back up and start them. I get nervous while turning on the wheelines because that is when something will go wrong. As the wheeline fills and the water pressure increases, a **riser** can blow or a **pipe latch** can unhook. The result is a mushroom cloud of water blowing up in the air, Dad driving across the field towards me at 40 miles an hour, and 30 more minutes of work. I know because I have made that mistake before!

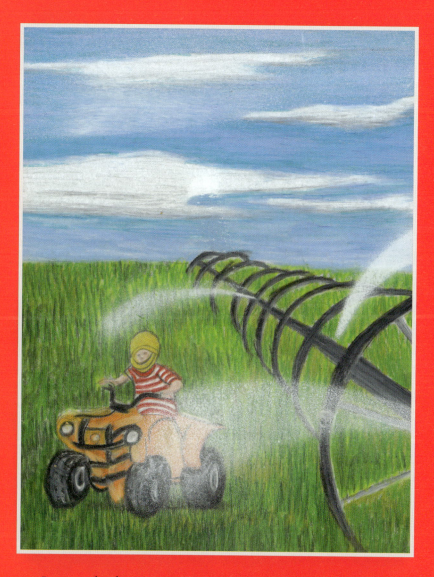

Later in the day I go out again and start the same process all over again. The humid heat is torture during the day and I feel like my clothes are wet from sweat. It is so hot that I get really frustrated while hooking the old pipes up and I just can't wait to be done! It feels great to be sprayed by the cold sprinklers after they have filled and there are no problems. After dinner I go out and change wheelines for the third time of the day. But, this time I go out and help irrigate other fields.

Wheat needs to be irrigated until the middle of July. I love watching the wheat turn from green to brown.

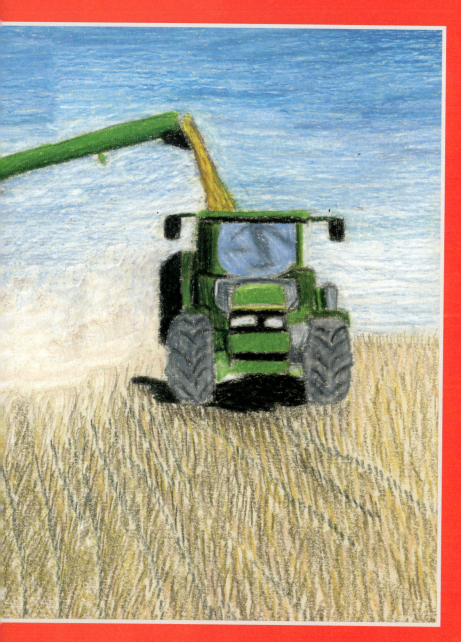

When the wheat heads fill and the crop starts to turn brown, Dad says, "We don't have to irrigate the field anymore." It's pretty tough. My uncle and dad harvest the wheat in the middle of August and put it in our **granary** for storage.

We grind the wheat we store into flour and sell it to bakeries and the local community. I help my dad run the wheat through the cleaner and I sit in there sometimes when he grinds the wheat into flour. The cleaner is a machine that separates the wheat seed from other seeds, small pieces of straw and wheat hulls.

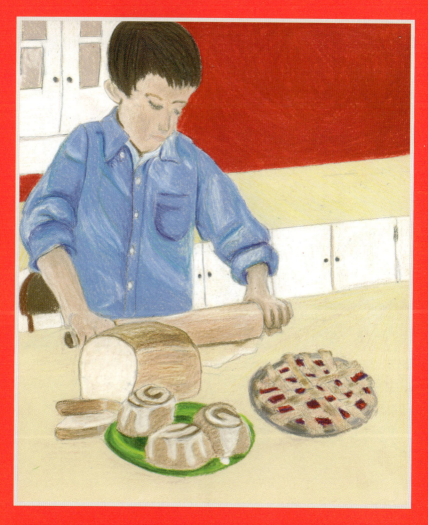

My favorite part though, is when we take some flour and I help my mom bake pies and bread. I love the homemade bread! Cinnamon bread is the best, it's so good it makes my mouth water just thinking about it. On our farm we enjoy watching the wheat grow from seeds planted in the dirt to pie crust that melts in our mouths. ❧

About the Author

Delaney Black, age 12

Growing up working on her family's farm not only inspired Delany Black with the idea for her award-winning story, *From Dirt to Pie Crust*, it also gave her the opportunity to share what she knows about agriculture with others. When given the assignment Delaney was excited to be able to write about her personal experiences on the farm. During her research, however, she also learned from her parents that she has five more years ahead of her working the wheelines! Delaney was surprised to learn how important agriculture is to California and looks forward to sharing more about it in future writings.

Delaney Black

About the Illustrators

Woodland High School - Woodland Joint Unified School District
Kristen Dachtler, Jennifer Long, and Lindsey Bratten
Art Instructor: Dawn Abbott

Kristen Dachtler, Jennifer Long and Lindsey Bratten from Woodland High School worked together as a team to illustrate the winning story, *From Dirt to Pie Crust*. The three friends were excited to have the opportunity to collaborate creatively on a unique school assignment. Although they live in a rural community where a few of their relatives own land, the three artists were all interested to learn about the process of both wheat growing and harvesting and enjoyed the challenge of bringing another student's words to life.

Kristen, Jennifer and Lindsey agree that this assignment taught them valuable life and career skills that they wouldn't normally have learned from a regular class project, including the importance of planning and communicating, working with a deadline and the value of group participation. They all look forward to seeing the finished product and being published illustrators.

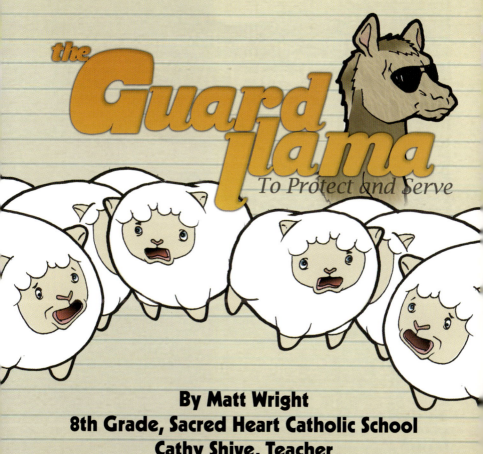

the Guard llama

To Protect and Serve

By Matt Wright
8th Grade, Sacred Heart Catholic School
Cathy Shive, Teacher
Stanislaus County
Illustrated by Inderkum High School

These days, coyotes are hungry and they love themselves a helping of sheep.

That's where I come in! Call me Guard Llama, that's my nickname, but my real name is Elliot Gawksalot. In case you don't know, I'm the defender of the sheep, a loyal soldier to the farmer, here to protect and serve the citizens of Double H Ranch.

All over California, guard llamas like me have been employed for the same reason, to protect the sheep at all costs.

This ground-breaking achievement has helped farmers all around the state have better **livestock** profits with even fewer **predators** killed.

Enough with the information, let's get down to business. I do a **perimeter** check every hour on the hour, just to make sure the citizens are safe and out of harm's way.

When I do encounter a predator threatening the livelihood of my citizens, I act fast with my special combat skills: screeching to disorient them, spitting to blind them, and kicking to just plain old get rid of them.

So far, my tactics have worked. The coyotes have been scared away, not even getting a whiff of my sheep. Well, all except one, Ace Bandit, the slyest, most evil coyote you have ever seen.

Ace Bandit and I go way back. We have been **rivals** ever since we first met, and now he threatens my sheep. He has one advantage, as I see it, and that is he never travels in a pack. The ragged carnivore always finds a way to get in the fence, always!

I must find a way to get rid of him. Or else the farmer will get rid of me! I won't let that happen. So let's do this. Bring all you've got Ace! You'll need it.

I heard him last night howling and he usually does that the night before he strikes. I have a plan. I have ordered all sheep to the middle of the **pasture** and I will guard the perimeter. It's time to bring Ace down! This will be an even fight: Guard Llama vs. Ace Bandit.

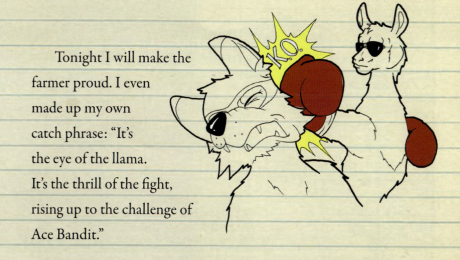

Tonight I will make the farmer proud. I even made up my own catch phrase: "It's the eye of the llama. It's the thrill of the fight, rising up to the challenge of Ace Bandit."

I have also rehearsed my combat tactics. First, I'll humiliate him in front of all the sheep. Then, we'll fight for a while, and finally, once I have won, I'll herd him to the state line. Yeah, Nevada sounds good. He can melt in the desert, for all I care.

Okay, it is dark now. I hear barking. The time has finally come for me to destroy Ace. There he is! Kick, spit, and scream! Kick, spit, and scream! This isn't going as I planned! He isn't snarling anymore, what happened? The sheep have made a circle around us now. Is he dead? "I knocked him out," said a familiar voice. "Let's get him on his way before he wakes up." We took him to the highway and put him on the back of a truck going to Nevada. I'm glad that's over with, I thought to myself.

When I got back to the pasture, I was surprised by the congratulatory party for me thrown by the farmer and the citizens. I talked with all of the sheep and before they went to bed, they thanked me for protecting them and I discovered that knocking out Ace was their way of returning the favor.

In the long run, you get what you give, and you should always treat others the way you want to be treated. I guess it just comes to me naturally. ❧

About the Author

Matt Wright, age 13

Matt Wright, an eighth grader at Sacred Heart Catholic School in Modesto, noticed there were a lot of llamas on the ranches in Stanislaus County and he grew interested in learning why. This sparked his imagination and became the subject of his award-winning story, *The Guard Llama: To Protect and Serve*. Matt used the contest as an opportunity to satisfy his own curiosity, while also applying critical thinking and creativity to craft fictional characters and a plot. He did some research and learned that having llamas can be a cost effective way to keep predators away from flocks of sheep and small livestock. When given the assignment, Matt was excited at the idea of winning a contest, and is thrilled that taking this chance has made him a published author!

About the Illustrator

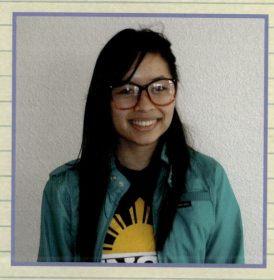

Inderkum High School - Natomas Unified School District

Leah Abucayan

Art Instructor: Michelle Townsend

High school senior Leah Abucayan, was excited to use the skills she learned from her digital photography and graphic design classes to illustrate *The Guard Llama: To Protect and Serve*. Leah hand drew each scene then transferred them into Photoshop where she was able to color them in and use techniques she learned in her class to complete the illustrations. Having some knowledge of agriculture from her biology class, Leah used that as a base for her research to make sure the illustrations were accurate. This story gave her a new understanding of how llamas are used in agriculture and a unique opportunity to illustrate another student's vision—a skill she will undoubtedly use in her future career aspiration in the industrial design field.

Glossary

Benefit
Something that promotes or enhances well-being.

Canal
A waterway used to bring water to a field to irrigate land.

Chutes
An inclined channel that guides flowing or falling materials to a designated location.

Clingstone
A fruit with flesh that sticks fast to the pit.

Dam
A barrier used across waterways to control the level of water.

Evaporate
The transformation of water from liquid to vapor.

Freestone
A fruit with flesh that does not stick to the pit.

Graded
Agriculture context: To be sorted and ranked by specific quality standards.

Granary
A building for storing grain.

Harvest
The process of gathering mature crops from the field.

Inspector
A person hired to check something for quality and standards.

Irrigation
An intentional application of water to the soil usually for assisting in the growth of crops.

Moisture
Wetness that can be felt in the atmosphere or condensed liquid on object surfaces.

Livestock
Domestic farm animals raised for production.

Orchards
Trees or shrubs that are planted and maintained for food production such as almonds and apples.

Organic
Farming according to certain standards, especially the exclusive use of naturally-produced fertilizers and alternative means of pest control.

Pasture
Grass or other vegetation eaten as food by grazing animals.

Perimeter
A border that surrounds an area.

Pipe latch
A simple latch used to connect pipes. A common way sprinkler pipes are put and held in place.

Predator
An animal which kills other animals for food.

Riser
A piece of machinery attached to a wheeline that casts out water when the tube is pressurized.

Rivals
Competitors; one who attempts to equal or exceed another.

Wheeline
An irrigation system that moves through a field under its own power.

Acknowledgments

California Foundation for
Agriculture in the Classroom

California Foundation for Agriculture in the Classroom (CFAITC) would like to acknowledge the many people who contributed to the success of the 2010 *Imagine this...* Story Writing Contest and the *Imagine this...* book.

Many thanks to...

Imagine this... Regional Coordinators
Sandra Gist-Langiano
Mary Landau
Doni Rosasco
Jacki Zediker

CFAITC Staff
Judy Culbertson, Publisher, *Imagine this...* book
Stephanie Etcheverria, Coordinator and Co-Editor,
     Imagine this... Story Writing Contest and book
Renee Hyatt, Co-Editor, *Imagine this...* book

CFAITC Board of Directors
Kenny Watkins, Chairman
Martha Deichler
Debbie Jacobsen
Jamie Johansson
Joe Peters
Rick Phillips
Jane Roberti
Craig Thomson

In partnership with

About California Foundation for Agriculture in the Classroom

The California Foundation for Agriculture in the Classroom (CFAITC) is a 501(c)(3) nonprofit organization that provides educators with quality free and low cost materials, training, and information to promote student understanding of California agriculture.

Agriculture in the Classroom is designed to help kindergarten through twelfth grade students acquire the knowledge necessary to become agriculturally literate. Through various programs and resources, educators are encouraged to incorporate agriculture into various subjects and lessons, and to point out the important role it plays in our economy and society.

Our mission is to increase awareness and understanding of agriculture among California's educators and students. Our vision is an appreciation of agriculture by all.

Why Agriculture?

Our world is a big, beautiful, intricate and often complicated puzzle. In this landscape, every piece plays a critical role, and each relies on another to make the image complete. Agriculture can help give perspective and meaning to the world around us through everyday examples such as plants and animals, demonstrating big-picture issues of water conservation, land use, the environment, and our food supply. Agriculture education creates unique and interactive opportunities to practice important life skills, such as teamwork and communication. Agriculture-based learning can also contribute to a child's increased confidence and improved attitudes and social skills in addition to making a positive impact in their own community.

Celebrating 25 Years

What began 25 years ago as one urban school district's plea to the California agriculture industry has grown into a statewide, far-reaching and reputable organization dedicated to educating the next generation of Californians and preserving our state's agricultural heritage. In the early 1980s, the San Francisco Unified School District was challenged to connect their city students to gain a basic understanding of their daily necessities and pleasures: the food in their lunchboxes, clothes on their back, pencil and paper used while sitting at their desks, the grass on their playing field—where did it all come from?

The California agriculture industry brought agriculture to life for thousands of San Francisco school children in May of 1981 by bringing livestock animals, farm equipment, and family farmers to 10,000 students on campuses across the city for San Francisco Farm Day. This event marked the impetus for the California Foundation for Agriculture in the Classroom.

A passionate group of individuals began California Foundation for Agriculture in the Classroom with small—but mighty—classroom resources and a calling to fill an important void within a growing number of California schools. In 25 years, our organization has touched the lives of more than 10,000,000 students!

Agriculture in the Classroom continues to develop resources, partner with like-minded organizations, recognize teachers and students who work as advocates of agriculture literacy and support the pursuit of agricultural careers to ensure the success of continuing our mission.

We look forward to another 25 years working to inform and inspire California students about the industry that sustains us all.

Imagine this...
Story Writing Contest
Annual Deadline: November 1

Attention Teachers! Use this creative writing contest in your classroom to meet California content standards in reading and writing for grades three through eight.

Students will discover how agriculture affects their everyday lives and their stories will inspire others to appreciate agriculture. State-winning students will have their stories illustrated by high school artists and published in next year's *Imagine this...* book.

Entry Details

- Submit up to five stories from each classroom
- Each story must have an entry form attached
- Stories must be written by one student author; no group entries allowed
- Stories exceeding 750 words will not be selected as state winners
- Stories should not be similar in theme to winning entries from previous years' contests

Winning stories must be

- Titled and original, creative student work
- Fact or fiction
- Appropriate for classroom use
- Grammatically correct
- Related to California agriculture in a positive way
- Typed, preferably, or neatly handwritten
- Written without reference to registered trademarks
- Suitable for illustration

Awards

Please refer to our website for a detailed list of the regional- and state-level awards.

www.LearnAboutAg.org

Entry Form

Send story and entry form to your regional coordinator (listed on back).

❏ Male ❏ Female

Student Name _____

Title of Story _____ Word Count _____

Teacher Name_____ Grade _____
(Print first and last name)

Teacher Signature _____ Class size _____

School Name _____

School Address_____

City, State, Zip _____

School Phone (   ) _____

County_____ Region Number (see map) _____

Teacher's E-mail _____

Principal_____
(Print first and last name)

Superintendent_____
(Print first and last name)

School District _____

How did you hear about this contest? _____

Hometown Newspaper _____

❏ Public School ❏ Private School ❏ Home School

Postmark by November 1, annually

Regional Coordinators

1 **Jacki Zediker**
7132 E. Louie Rd.
Montague, CA 96064
(530) 459-0529

Butte	Nevada
Colusa	Plumas
Del Norte	Shasta
Glenn	Sierra
Humboldt	Siskiyou
Lake	Sutter
Lassen	Tehama
Mendocino	Trinity
Modoc	Yuba

Send story and entry form, postmarked by Nov. 1, to your regional coordinator.

2 **Doni Rosasco**
16002 Hwy. 108, Jamestown, CA 95327
(209) 984-3539

Alameda	El Dorado	San Francisco	Solano
Alpine	Marin	San Joaquin	Sonoma
Amador	Napa	San Mateo	Stanislaus
Calaveras	Placer	Santa Clara	Tuolumne
Contra Costa	Sacramento	Santa Cruz	Yolo

3 **Sandi Gist-Langiano**
P.O. Box 748
Visalia, CA 93279
(559) 732-8301

Fresno	Merced
Inyo	Mono
Kern	Monterey
Kings	San Benito
Madera	San Luis Obispo
Mariposa	Tulare

4 **Mary Landau**
330 East Las Flores Drive
Altadena, CA 91001
(626) 794-4025

Imperial	San Bernardino
Los Angeles	San Diego
Orange	Santa Barbara
Riverside	Ventura